HATCH!

Written by Cassie Hoyt

Illustrated by Amanda Crawford Brown

For my sweet and curious boys, Lincoln and Connor. I love you!
-CH

For those who gave me support and gave me a chance! Tracy, Todd G, Alicen, and of course, Cassie for trusting me with your amazing little egg book!
-ALCB

Cassie Hoyt • Greenville, NC
www.cassiehoyt.com

contact: info@cassiehoyt.com

Cover by Amanda Crawford Brown.

The egg is

HATCHING!

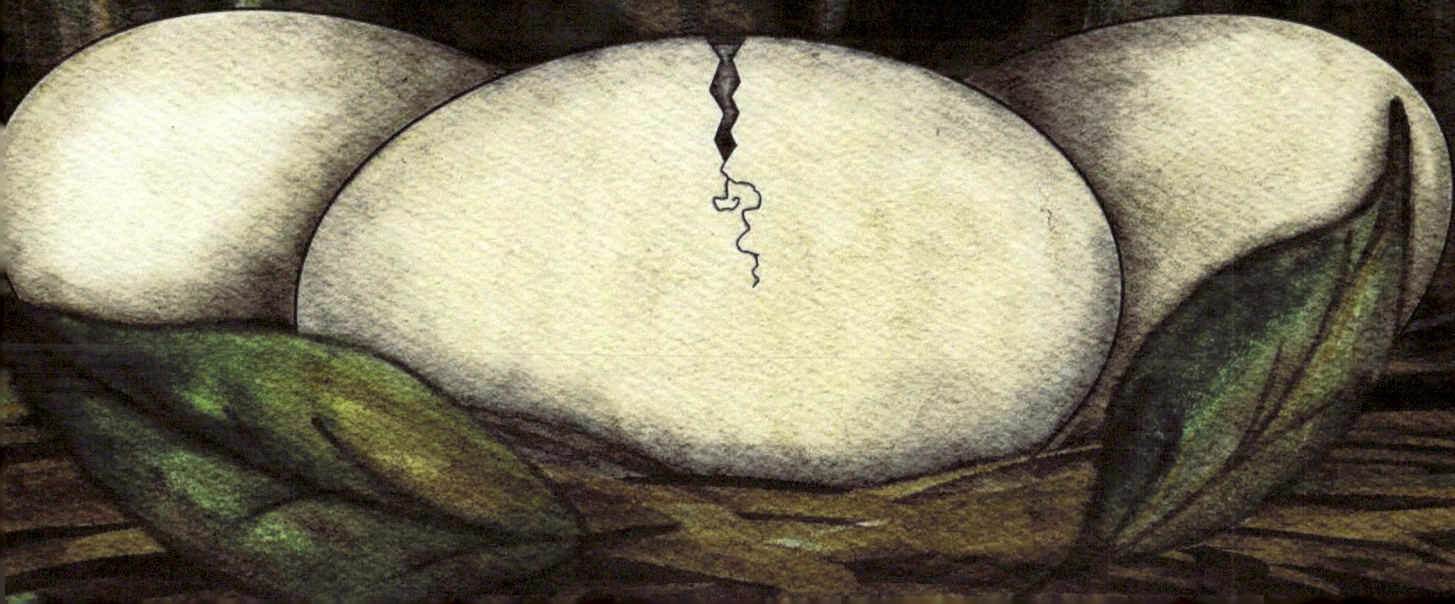

WHAT WILL WE SEE?

It's a
LiTTLE
DUCKLiNG

Staring at **ME!**

The egg is
HATCHING!

It's a
BABY
ALLiGATOR

The egg is HATCHING!

WHAT WILL WE SEE?

It's a SLIMY SNAKELET

Staring at ME!

The egg is
HATCHING!

WHAT WiLL WE SEE?

It's a
BABY BiRD

Staring at
ME!

The egg is HATCHING!

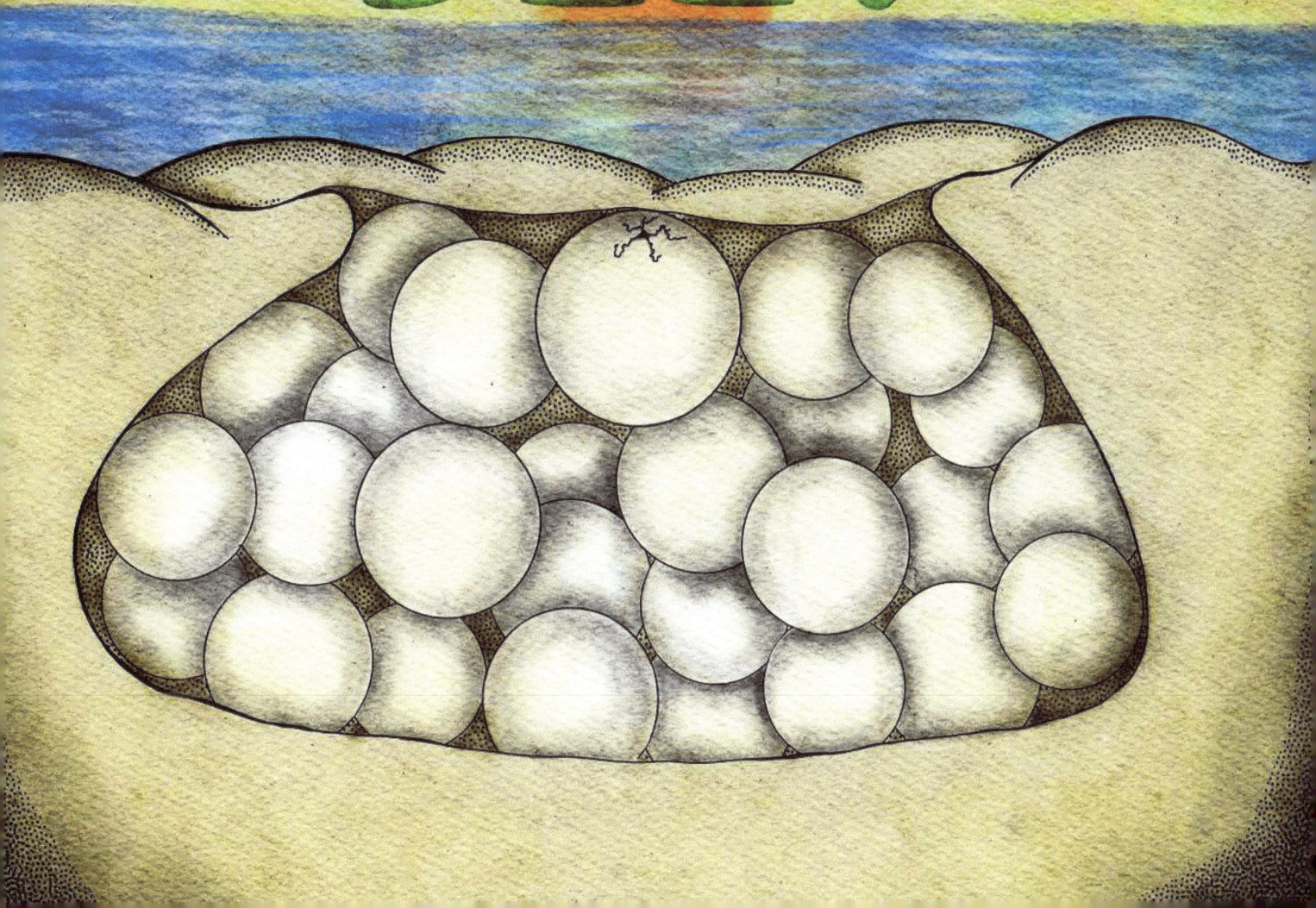

WHAT WILL WE SEE?

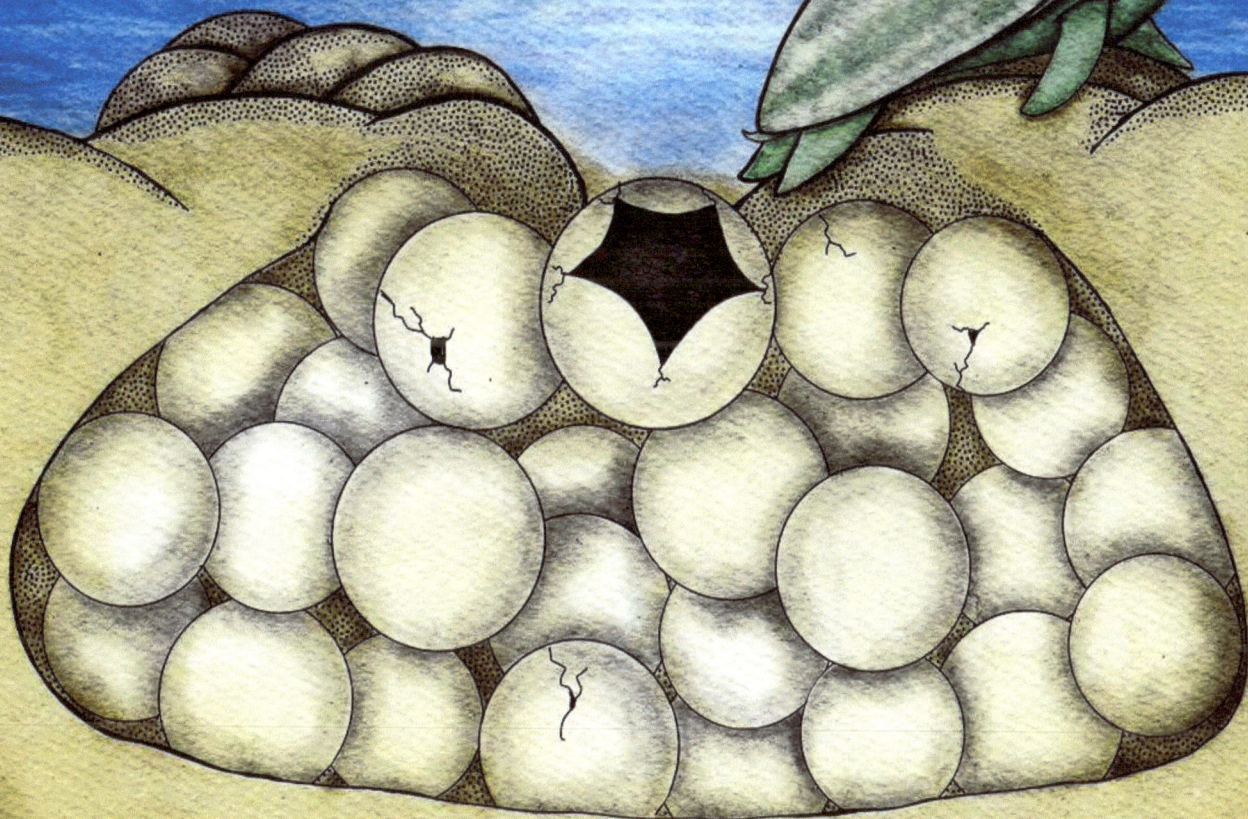

The egg is
HATCHING!

It's a
BABY
PENGUIN

Staring at
ME!

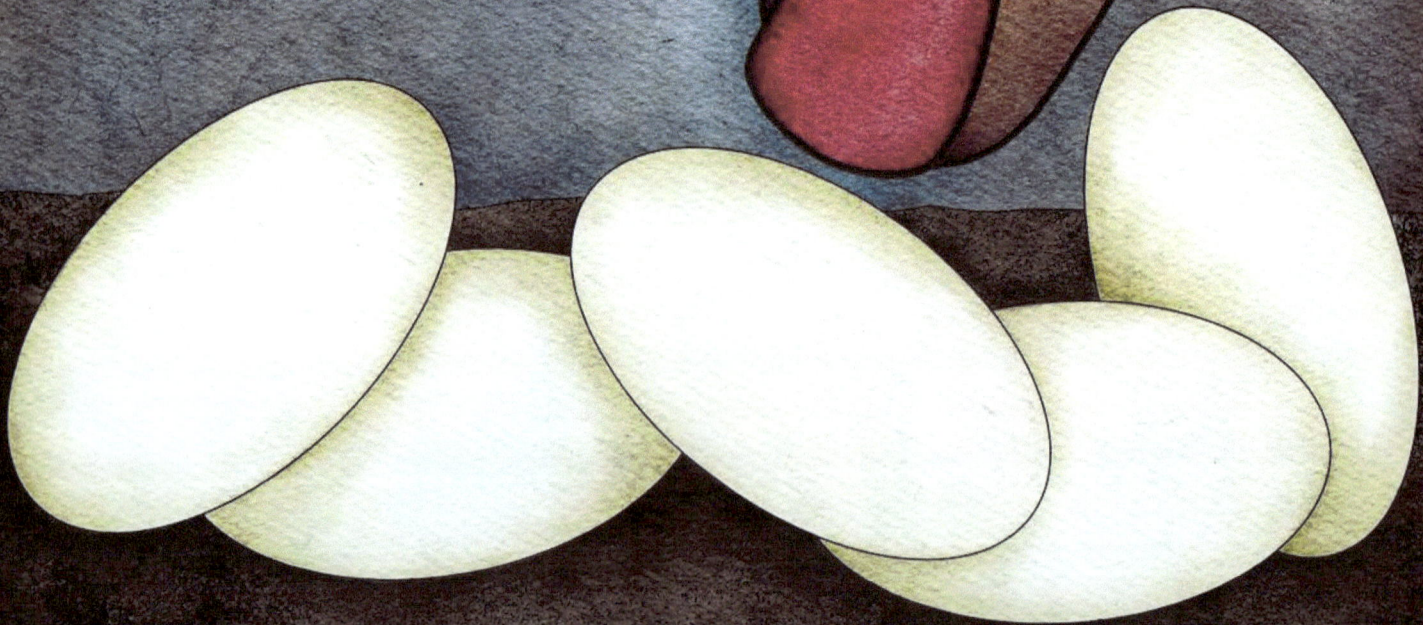
The egg is HATCHING!

It's a
TiNY
LiZARD

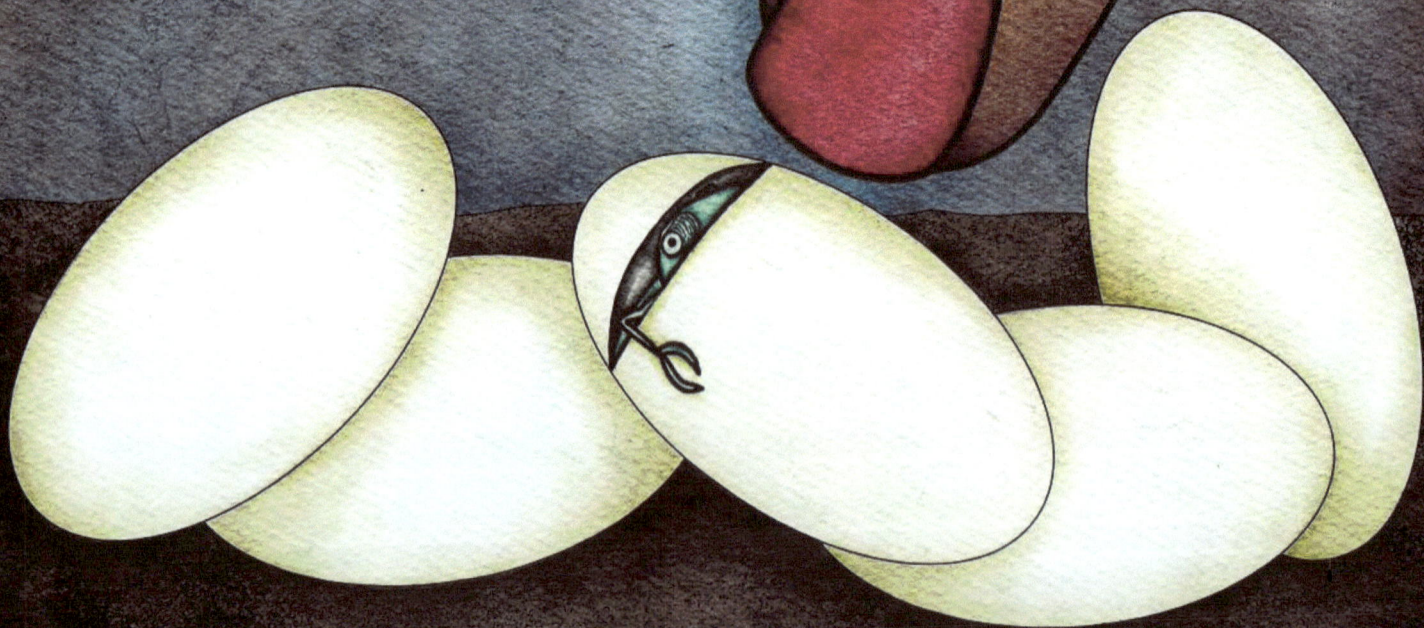

Staring at ME!

The egg is
HATCHING!

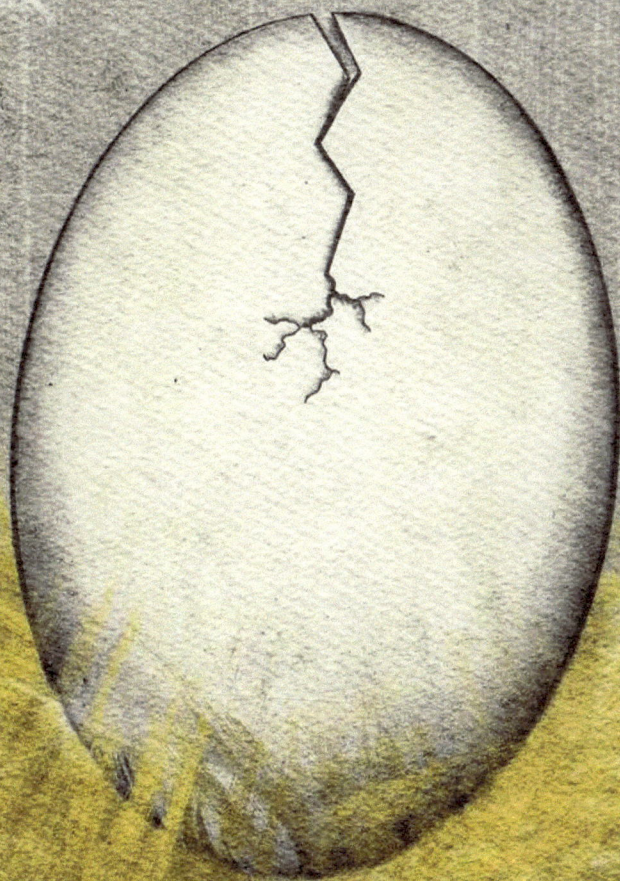

WHAT WILL WE SEE?

It's a
BABY CHICK

Staring at **ME!**

The egg is

HATCHiNG!

WHAT WiLL WE SEE?

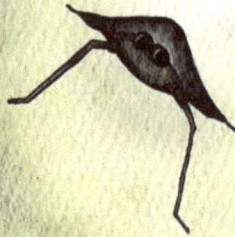

It's a **TEENY SPiDERLiNG**

Staring at

ME!

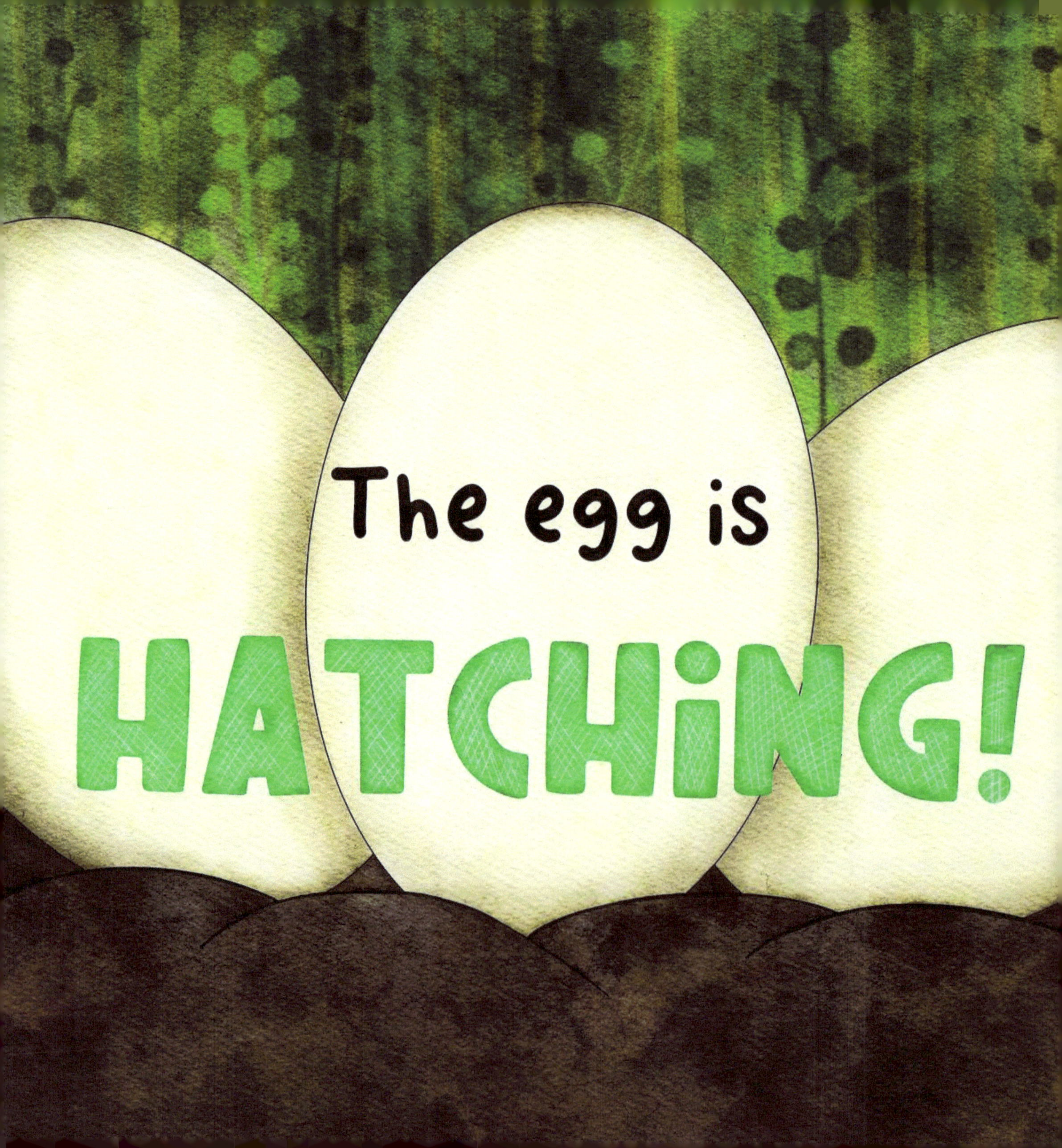

The egg is

HATCHiNG!

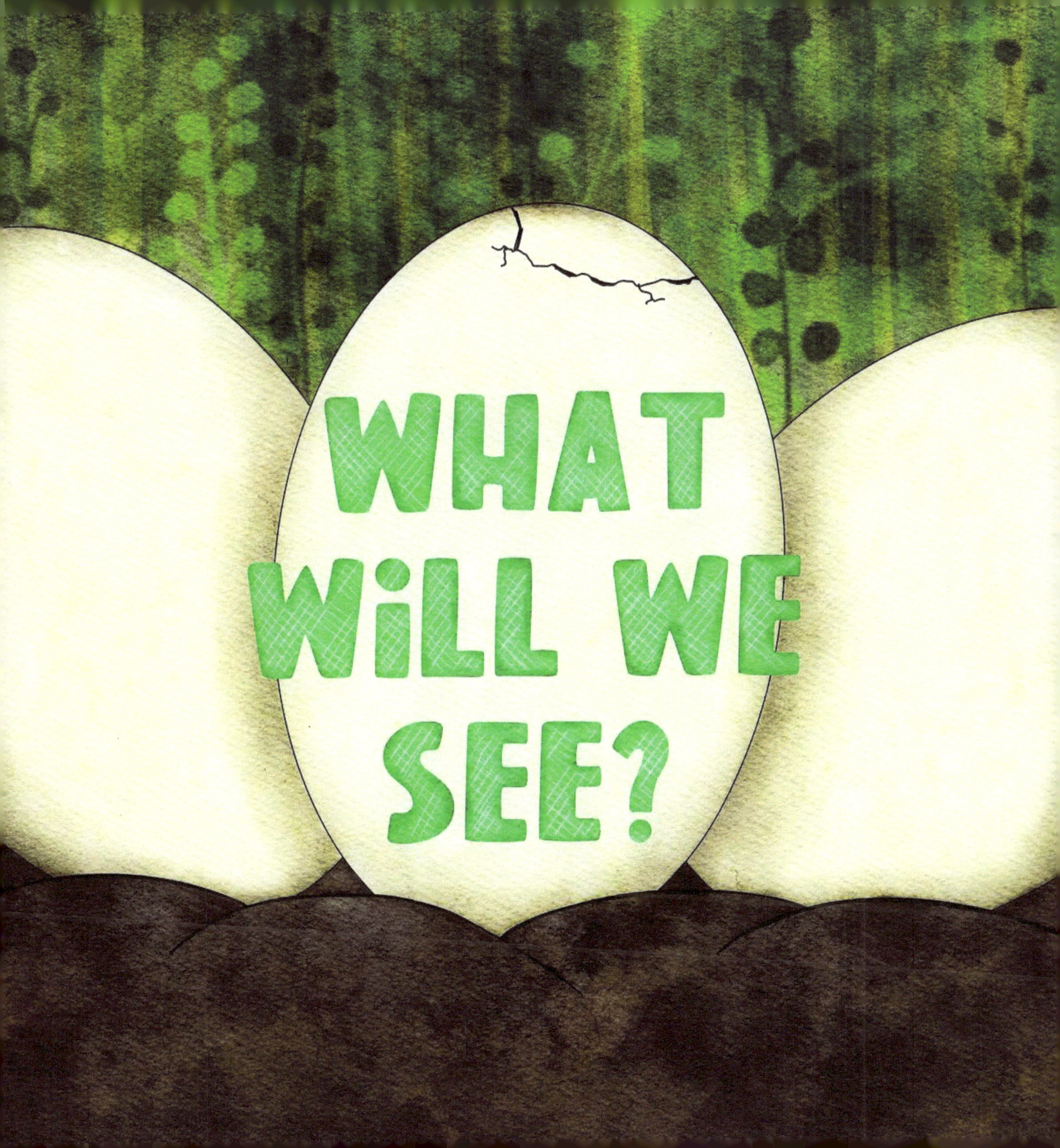
WHAT
WiLL WE
SEE?

It's a
BABY
OSTRiCH

The egg is

HATCHING!

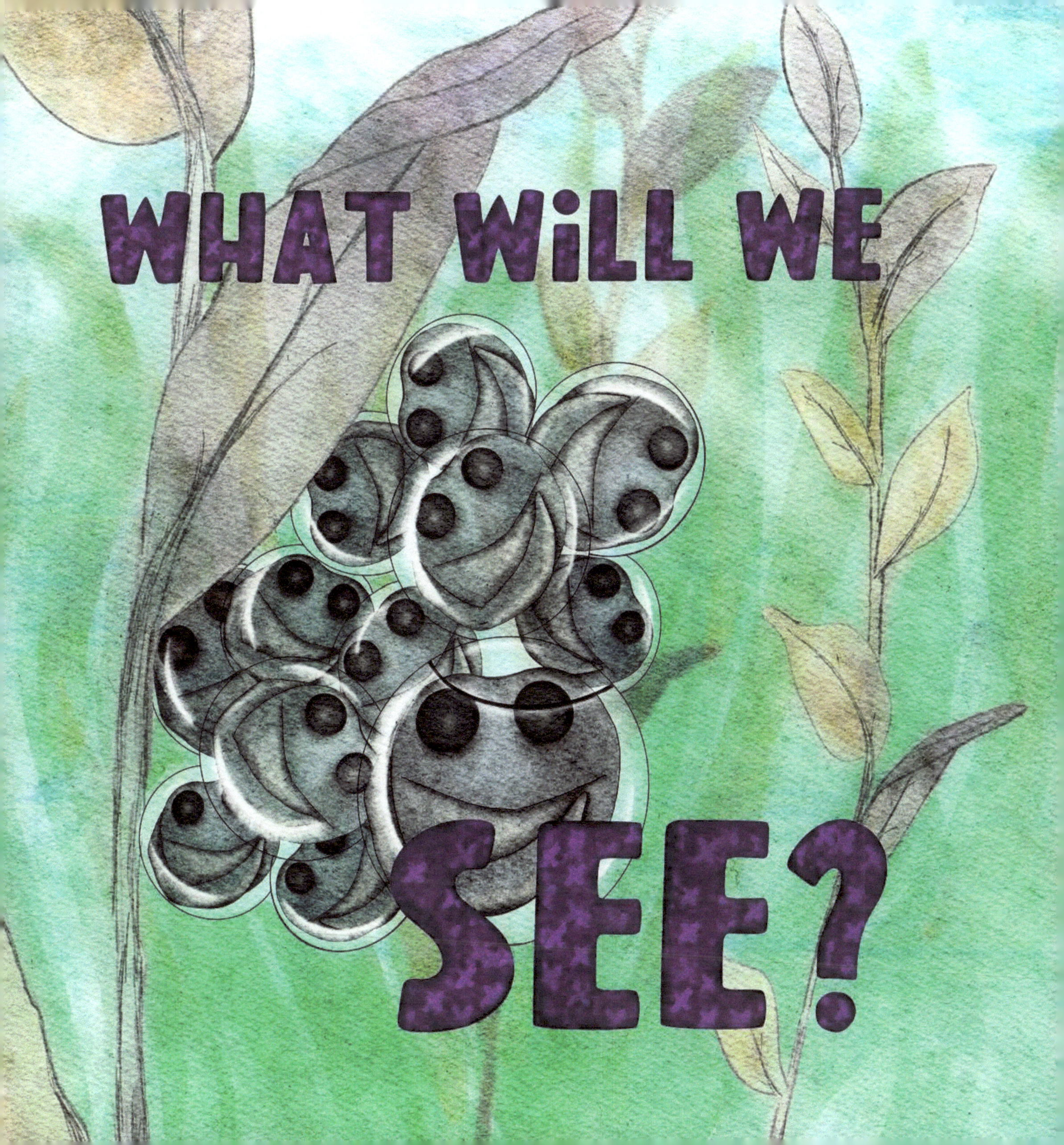

WHAT WiLL WE

SEE?

It's a

LiTTLE FROG

THE

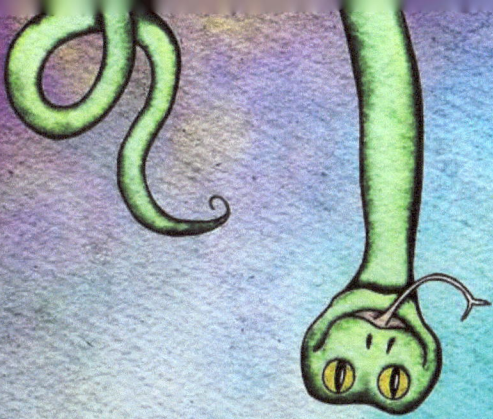

www.ingramcontent.com/pod-product-compliance
Lightning Source LLC
Chambersburg PA
CBHW060843270326
41933CB00003B/182